# WORLD BOOK

OUR SOLAR SYSTEM

Jupiter
AND
Saturn

THE GAS GIANTS

Contents

- 4 Jupiter: King of the Planets
- 6 Jupiter: Fifth Planet from the Sun
- 8 Stripes and Swirls
- 10 A World of Gas
- 12 Jupiter and Earth
- 14 Weather on Jupiter
- 16 Jupiter's Great Red Spot
- 18 A Year and a Day on Jupiter
- 20 Jupiter's Interior
- 22 Jupiter's Moons
- 24 The Galilean Moons
- 26 Io
- 28 Europa
- 30 Jupiter Has Rings!
- 32 Exploring Jupiter
- 34 Can There Be Life on Jupiter?

Saturn

- 36 Beautiful Saturn
- 38 Saturn: Vital Info
- 40 Saturn and Earth
- 42 Saturn's Structure
- 44 Saturn's Atmosphere
- 46 Saturn's Weird Weather
- 48 The Rings of Saturn
- 50 Saturn's Disappearing Rings
- 52 Exploring Saturn
- 54 Saturn's Moons
- 56 Mysterious Titan
- 58 Enceladus
- 60 Glossary
- 62 Index
- 64 Acknowledgments

If a word is printed **in bold letters that look like this** the first time it appears on any page, you will find the word's meaning in the glossary beginning on page 60.

Astronomers use different kinds of photos to learn about such objects in space as planets. Many photos show an object's natural color. Other photos add false colors or show types of light that the human eye cannot normally see. When appropriate, the captions in this book will state whether a photo uses false color. Other photos and illustrations use color to highlight certain features of interest.

Jupiter
King of the Planets

Magnificent Jupiter is king of the **planets!** Jupiter, the largest planet in our **solar system,** is one of the brightest objects in the night sky. Because it is so prominent, it is named after the king of the gods and ruler of the universe in Roman **mythology.** He was also called *Jove*.

Astronomers call the four largest planets in the solar system the *gas giants*. These planets are Jupiter, Saturn, Uranus, and Neptune. They have earned that name for good reason—all are huge planets that are made up mainly of gas.

Dozens of **moons** and four faint rings surround Jupiter. Astronomers sometimes refer to the planet and its moons and rings as the *Jovian system*.

Jupiter has the most **mass** compared to its neighbors. Mass is the amount of matter a thing contains. Jupiter has **twice as much mass** as all of the other planets combined!

This stunning image of Jupiter shows the beautiful stripes and swirls seen on the largest planet in our solar system.

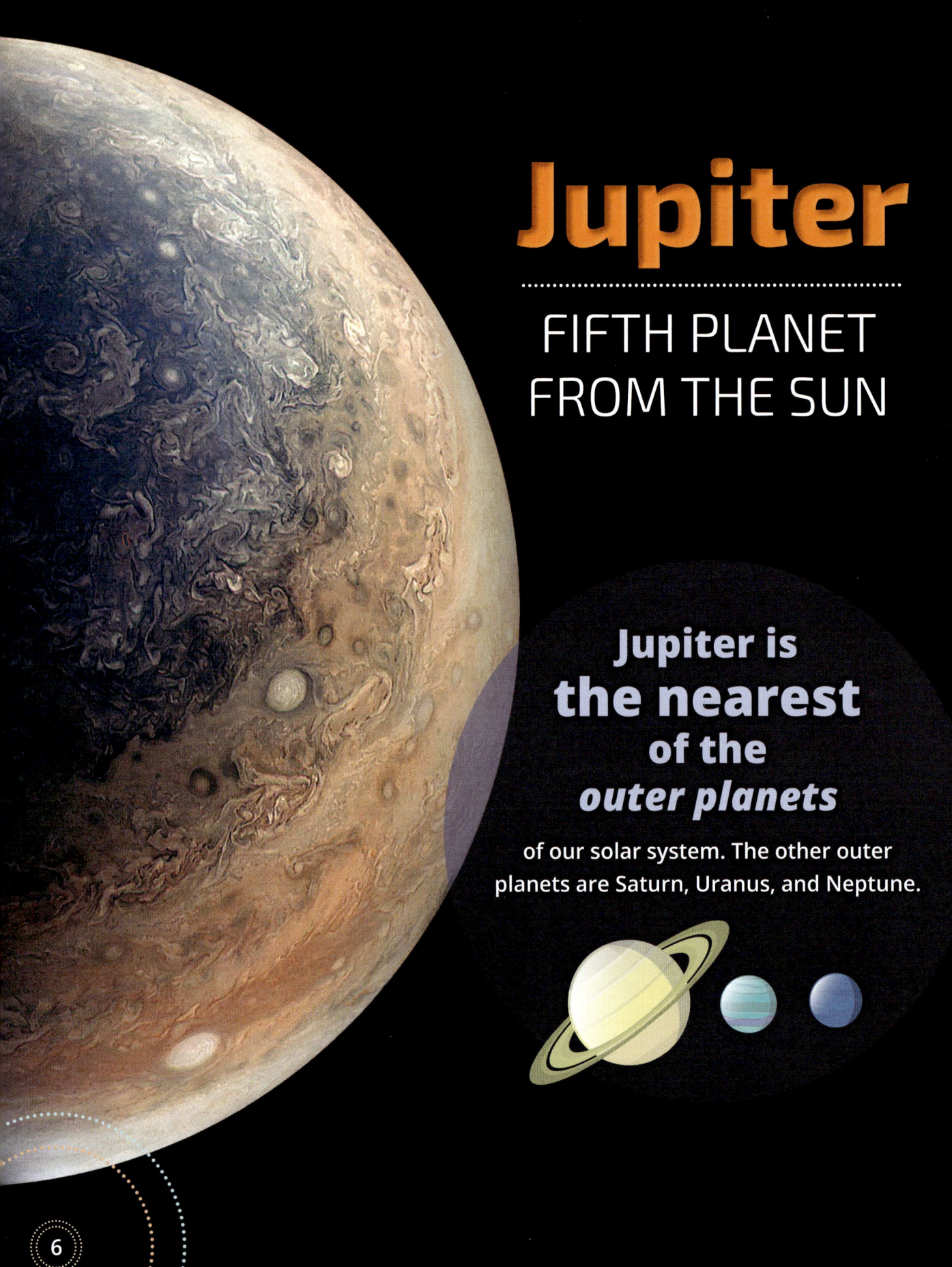

Jupiter

FIFTH PLANET FROM THE SUN

Jupiter is **the nearest** of the *outer planets* of our solar system. The other outer planets are Saturn, Uranus, and Neptune.

Jupiter's distance from the sun changes over time because the planet follows an *elliptical* (oval-shaped) **orbit.**

On average, Jupiter's orbit lies about 484 million miles (779 million kilometers) from the sun, **between the orbits of the planets Mars and Saturn.**

Jupiter is really far away from Earth!

At their closest distance, the two planets are about 366 million miles (589 million kilometers) apart. Even though Jupiter is about five times as far from the sun as Earth, it is easily visible in the night sky.

Jupiter formed closer to the sun than where it orbits today!

It may have then moved even closer before spiraling back to its current position about 4 billion years ago.

It takes almost **44 minutes** for light from the sun to reach Jupiter!

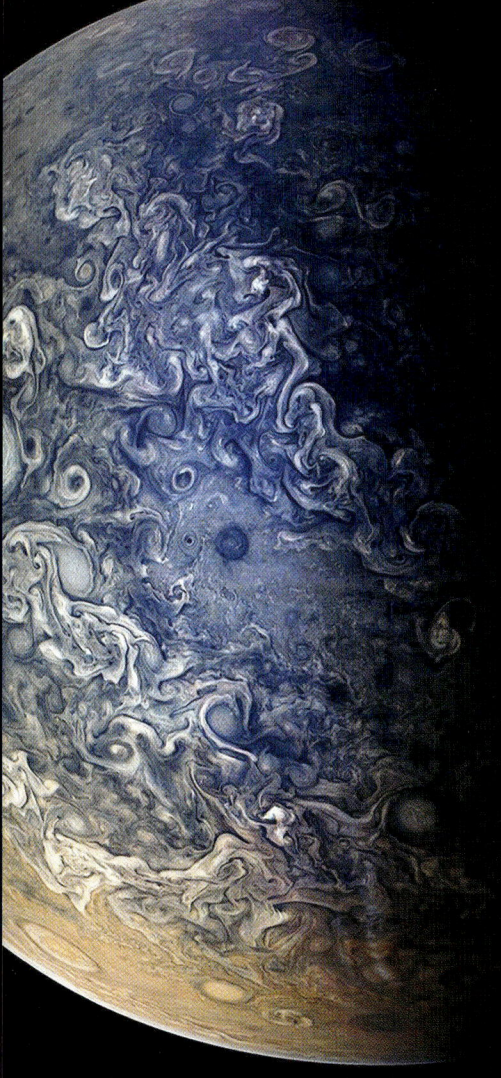

Little Red Spot

Stripes and Swirls

Jupiter's most visible features are broad stripes of light and dark clouds. They range from orange-brown to bluish-white. Jupiter's **atmosphere** also has many swirls or circular features. A gigantic red spot, larger than Earth, is visible on the southern half of the planet. Scientists call this feature the Great Red Spot (page 16).

The Great Red Spot is a huge storm

that has raged for hundreds of years. In the early 2000's, three storms on Jupiter combined to form a single large swirl. This new spot is now called the Little Red Spot.

Striking cloud patterns on Jupiter are seen in this series of images taken from space. A storm known as the Little Red Spot is visible near the bottom of the second and third images.

A World of Gas

Outer layer of mostly hydrogen gas

Liquid metallic hydrogen

Core

As a gas giant planet, Jupiter does not have a surface like Earth does. Jupiter is mainly a huge ball of gas.

Jupiter's outer layer is made chiefly of clouds of **hydrogen** gas. It also contains small amounts of **helium** and a few other chemical elements. Most of the elements in Jupiter's atmosphere are combined into **molecules** of water and **ammonia.**

Below the clouds is a layer of liquid hydrogen. The enormous pressure of the layer above causes the hydrogen **atoms** there to squeeze together and become a liquid. About 6,000 miles (10,000 kilometers) below the clouds, the liquid hydrogen turns into liquid **metallic** hydrogen. This unusual form of hydrogen does not exist on Earth.

Clouds give Jupiter its colorful, banded appearance. The clouds are made up mainly of icy particles of ammonia and other chemicals.

Astronomers think Jupiter may have a solid **core** at its center. The core is probably about the size of Earth.

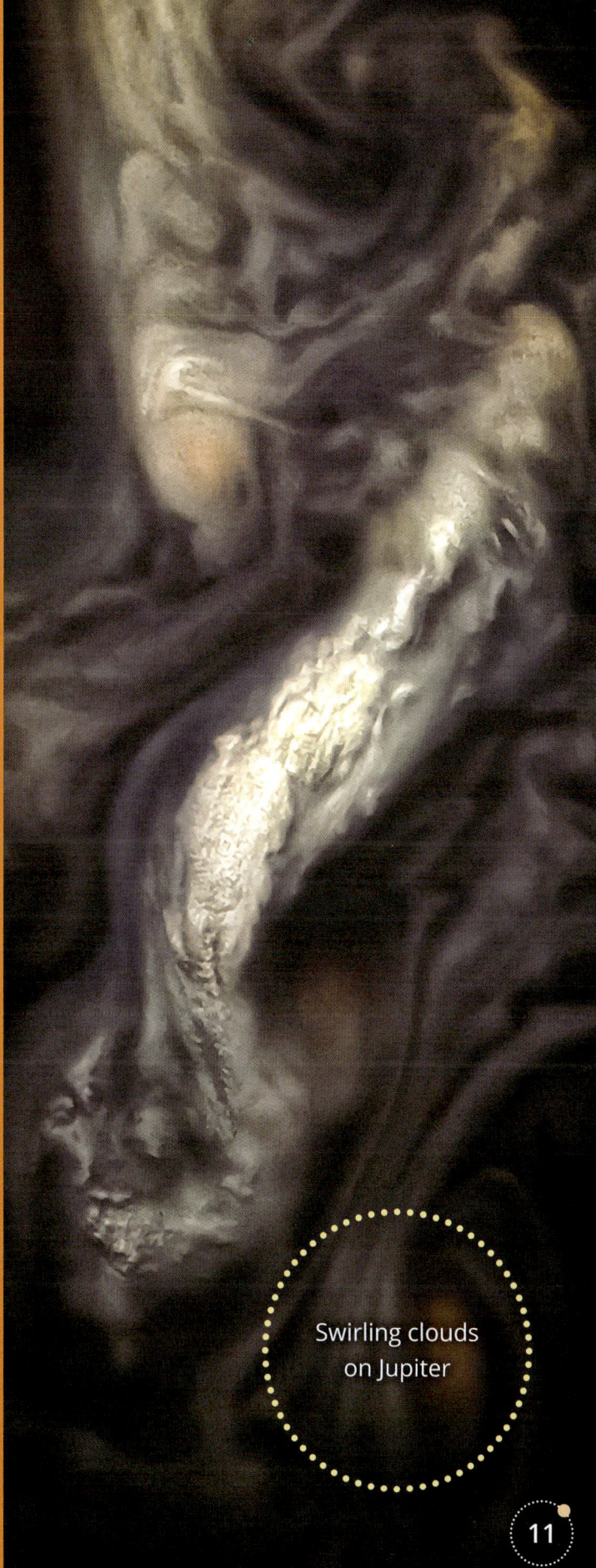

Swirling clouds on Jupiter

Jupiter and Earth

A COMPARISON

Length of day on Jupiter:
9 hours 55 minutes

Jupiter has **hundreds of times** the mass of Earth. But Jupiter's average *density* is much lower than Earth's. (Density is the amount of matter in a given space.) In fact, Jupiter is only a bit more dense than liquid water on Earth.

Jupiter's density is so low because it is made mostly of the **two lightest elements:** hydrogen and helium gases. Earth is a rocky planet made of mostly heavier materials.

240

Because Jupiter has much more mass than Earth, its **gravity** is much stronger. Jupiter's gravity is about

2 ½ times

as strong as that of Earth. If you weighed 100 pounds (45 kilograms) on Earth, you would weigh about 240 pounds on Jupiter (109 kilograms).

Jupiter is about

11 times

the *diameter* (distance across) of our own planet Earth! Its diameter at the equator measures 88,846 miles (142,984 kilometers). In fact, Jupiter takes up so much space that if it were hollow, more than 1,300 Earths would fit inside!

Jupiter has at least

79

moons—
and counting!

Weather on Jupiter

An artist's illustration of the cloud tops on Jupiter

The weather on Jupiter is always cloudy and windy. The winds on Jupiter blow constantly and can reach speeds of about 400 miles (650 kilometers) per hour near the **equator.** That's almost twice as fast as the most destructive hurricane winds on Earth. Lightning bolts more powerful than any on Earth flash in Jupiter's atmosphere.

Temperatures on Jupiter vary widely by height and location. Near Jupiter's cloud tops, the temperature is roughly –236 °F (–149 °C). In contrast, the core of the planet may be up to 43,000 °F (24,000 °C)—that's hotter than the surface of the sun!

Scientists think that temperature differences power Jupiter's winds. This process on Jupiter can be seen from space. The pale bands of clouds on Jupiter are areas where the temperature is warmer and the clouds and gases are rising. The darker bands are cooler areas where the clouds and gases are sinking.

This image shows Jupiter's southern half from space. The oval features are huge storms, up to 600 miles (1,000 kilometers) across.

Jupiter's Great Red Spot

Jupiter's Great Red Spot is actually a huge storm that has been raging for hundreds of years!

It has been visible from Earth since at least 1831. But scientists have observed over many years that the Great Red Spot is slowly shrinking. One day, it may disappear completely!

The Great Red Spot is wider than the diameter of Earth!

Jupiter's Great Red Spot is so large that it can be seen through a **telescope** from Earth.

A Year and a Day on Jupiter

Jupiter takes a long time to complete one orbit around the sun. Because it is so far away from the sun, Jupiter takes almost 12 Earth **years** to make a complete orbit. So, one year on Jupiter is equal to 12 years on Earth!

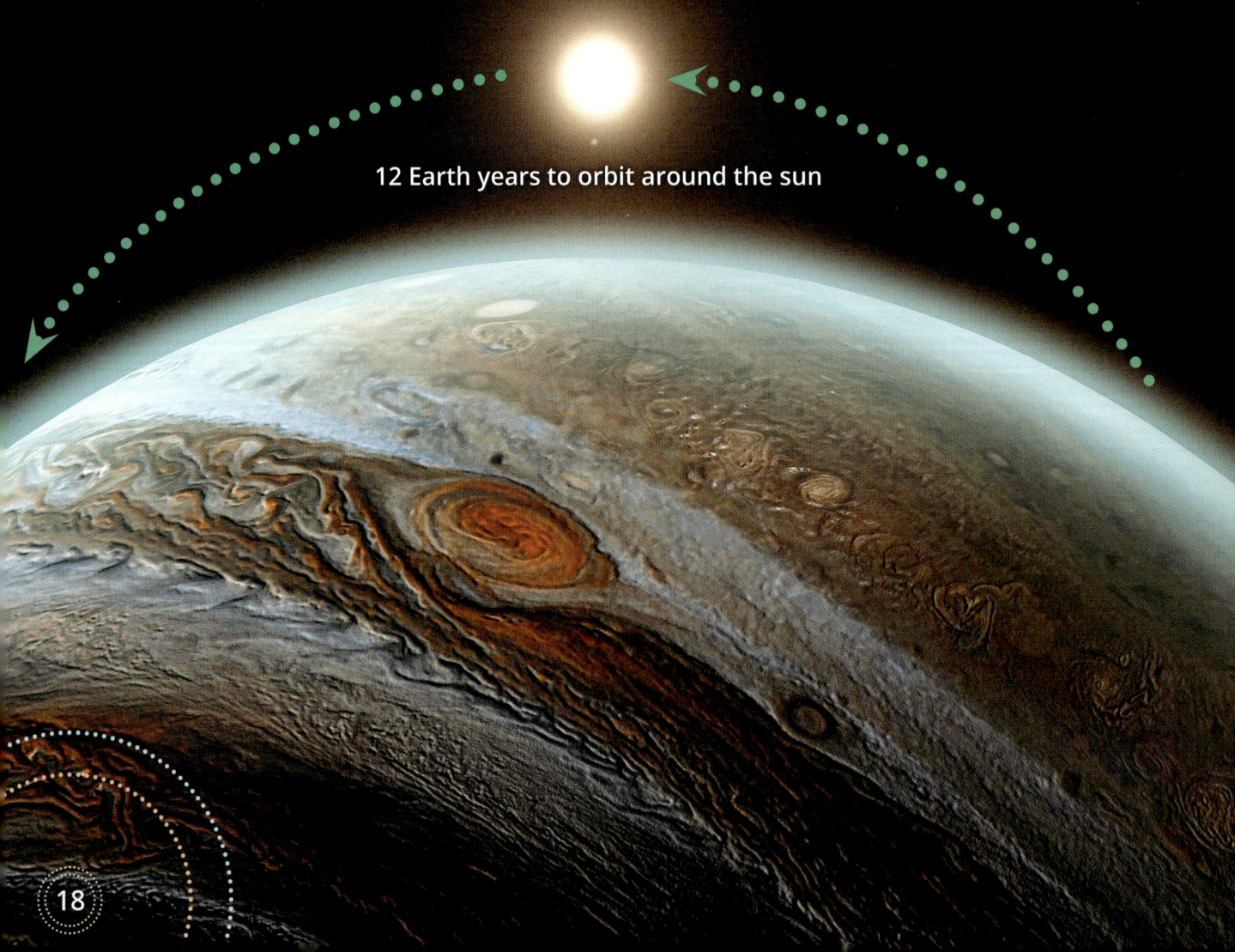

12 Earth years to orbit around the sun

Magnetic Planet

The metallic hydrogen around Jupiter's core gives the planet the **strongest magnetic field in the solar system.** This magnetic field creates a dangerous radiation belt around Jupiter that can damage visiting spacecraft.

Jupiter may have a long year, but it has the shortest **day** of any planet in our solar system! Earth makes a complete rotation on its **axis** about every 24 hours. Jupiter spins around much faster. A day on Jupiter lasts slightly less than 10 hours.

Because Jupiter rotates so quickly, the planet's shape is not perfectly round. Instead, Jupiter bulges slightly along the equator.

Jupiter's magnetic field creates **auroras** when high-energy particles enter the planet's atmosphere near its poles and collide with atoms of gas. An **aurora** is a colorful glow that appears in the sky at night.

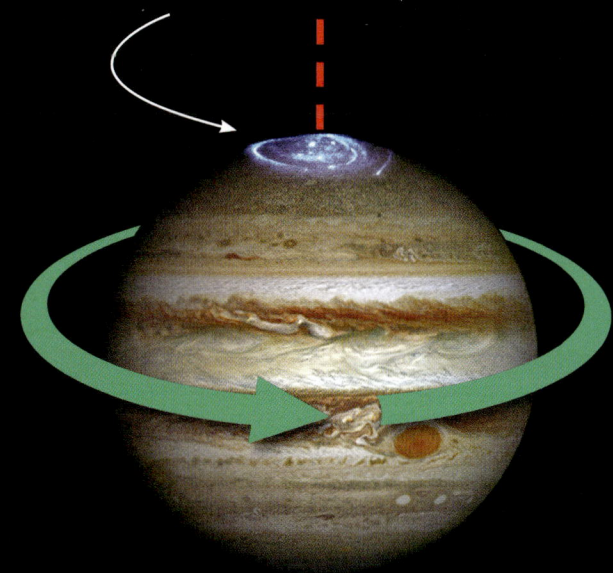

Jupiter's Interior

To an observer from space, Jupiter's cloud tops may look like a solid surface. But no one could stand on the surface of Jupiter. Although Jupiter has a very thick atmosphere, the clouds and gases are not thick enough to support people standing on top of them. If Jupiter did have a surface, that surface would be very cold and windy.

Jupiter does not receive nearly as much heat from the sun as Earth does. However, Jupiter *radiates* (gives off) nearly twice as much

heat as it absorbs from the sun. Scientists believe that some of this heat is energy left over from Jupiter's formation as a planet. Some of the energy might come from heat created as Jupiter slowly shrinks under its own gravity.

An artist's illustration of the cloud tops in Jupiter's atmosphere. Lightning is seen below. Two of the Jovian moons can be seen in the sky.

Jupiter's Moons

Jupiter has at least 79 moons! Jupiter's moons vary widely in size, color, atmosphere, and density.

Fifty-three of Jupiter's moons have been given names. The names come from Greek or Roman mythology. Most of these moons range from 10 to 104 miles (16 to 167 kilometers) in diameter.

Astronomers continue to discover other tiny moons orbiting Jupiter. Some of these moons are less than 1.2 miles (2 kilometers) in diameter. The tiny moons are not given names until their discovery is confirmed. Scientists think Jupiter's smallest moons have yet to be discovered.

Because so many moons orbit Jupiter, astronomers sometimes describe Jupiter and its moons as a mini solar system!

FUN FACT

Three of Jupiter's moons—Io, Ganymede, and Callisto—are all **bigger than Earth's moon.**

This "family portrait" image of the Jovian system shows the edge of Jupiter with Jupiter's four largest moons, known as the Galilean satellites.

The Galilean Moons

Four of Jupiter's moons are so large that they are visible from Earth to observers using a simple telescope. These moons—named Io *(EYE oh)*, Europa, Ganymede *(GAN uh meed)*, and Callisto *(kuh LIHS toh)*—are also called the Galilean moons. They are called this because the Italian astronomer Galileo *(gal uh LAY oh)* first observed these moons, in 1610.

Ganymede is the largest of Jupiter's moons. It is also the largest moon in the solar system—larger than the planet Mercury. Callisto is the farthest Galilean moon from Jupiter. It is almost completely covered in craters.

Besides the Galilean moons, other moons of Jupiter are arranged into two groups called the inner **satellites** and the outer satellites. The inner satellites are closer to Jupiter than the Galilean moons. The outer satellites are farther from Jupiter than the Galilean moons.

The discovery of moons orbiting another planet helped convince Galileo and others that

Earth was not at the center of the universe.

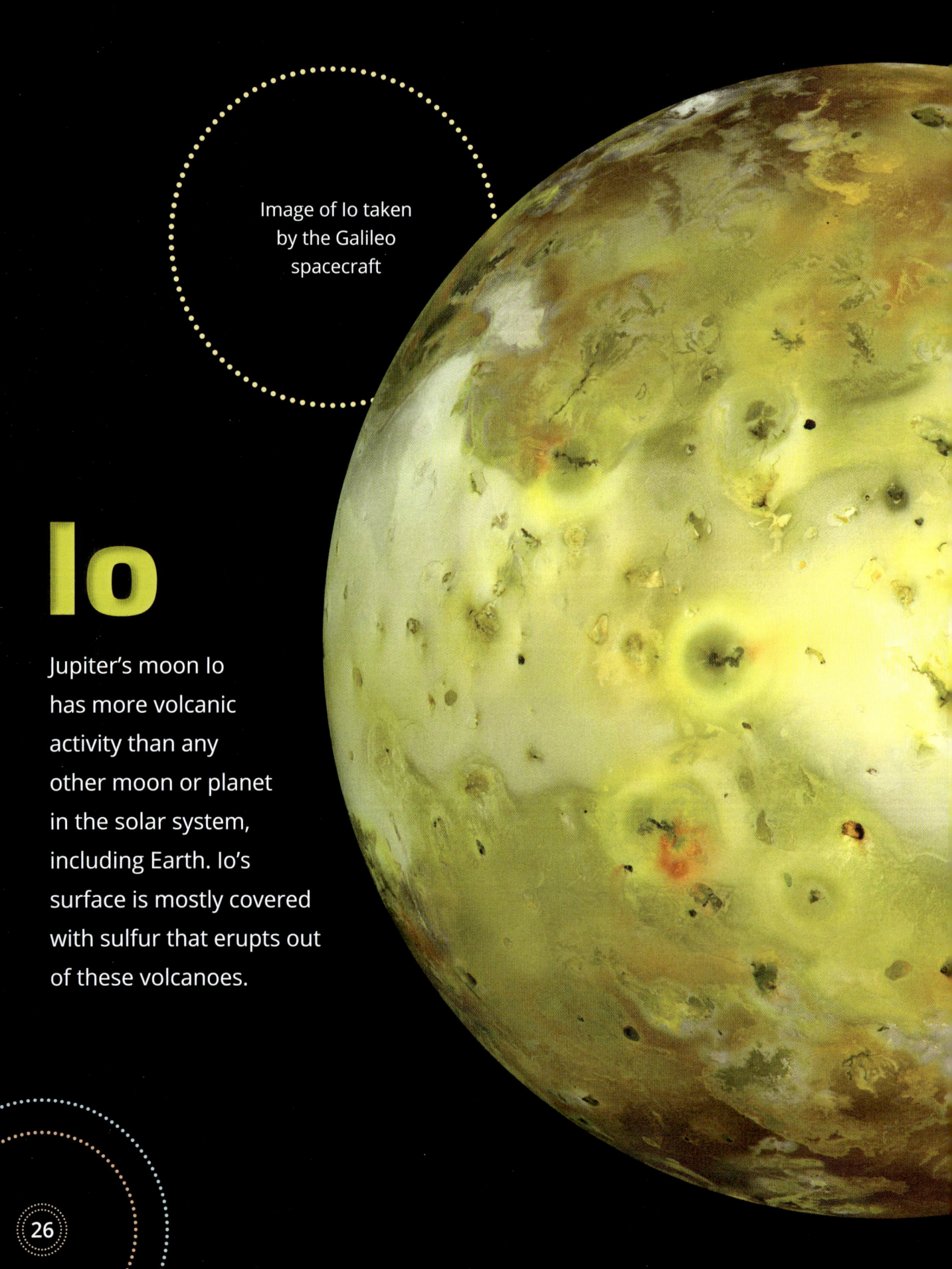

Image of Io taken by the Galileo spacecraft

Io

Jupiter's moon Io has more volcanic activity than any other moon or planet in the solar system, including Earth. Io's surface is mostly covered with sulfur that erupts out of these volcanoes.

As Io travels in its orbit around Jupiter, the gas giant's powerful gravity causes the solid surface of Io to rise and fall around 300 feet (100 meters). This constant flexing and bending generates heat that powers Io's volcanoes. The volcanoes spew sulfur and other materials that coat the surface of the moon.

FUN FACT

Volcanic eruptions on Io **have been seen from space!**

Europa

The surface of Jupiter's moon Europa is mostly **water ice.** This surface has many cracks, valleys, and ridges. Many scientists think that beneath the moon's icy crust there is a deep ocean of liquid water or slushy water ice.

Scientists think Europa is one of just a few places in our solar system that could possibly harbor life. If living things do exist in the ocean underneath Europa's surface, they would likely be very different from living things on Earth. Scientists are designing new space **probes** to visit Europa to find out if life exists there!

The fascinating surface of Jupiter's frozen moon Europa looms large in this image. Long cracks and ridges crisscross the surface ice, interrupted by regions where the ice has been broken up and re-frozen into new patterns.

An artist's view of the Voyager spacecraft near Jupiter

Jupiter Has Rings!

Astronomers did not discover that Jupiter had rings until 1979. That year, two space probes from Earth called Voyager 1 and Voyager 2 flew by Jupiter. During their flybys, the Voyager probes took pictures of what appeared to be at least two rings.

Additional pictures taken by another probe during the 1990's confirmed that Jupiter has four faint rings. Jupiter's rings are much fainter than the rings around Saturn.

Jupiter's brightest ring is called the *main ring*. A fainter ring is called the *halo ring*, and two even fainter rings are called the *gossamer rings*.

Dust and Ice

Astronomers think that Jupiter's rings are made of tiny bits of dust. The dust is knocked off Jupiter's inner moons when they are struck by meteoroids. The edges of the rings are shaped by the orbits of Jupiter's four inner satellites.

Exploring Jupiter

In 1973, **Pioneer 10** was launched by the United States National Aeronautics and Space Administration (NASA). It became the first spacecraft to fly past Jupiter. In 1974, **Pioneer 11** studied Jupiter and Saturn. These probes sent back information about Jupiter's atmosphere, gravity, and magnetic field.

In 1979, the NASA probes **Voyager 1** and **Voyager 2** flew past Jupiter. These probes studied Jupiter's atmosphere, discovered its rings, and photographed many of its moons.

Ancient astronomers knew about Jupiter because the planet can easily be seen with the unaided eye. Today, astronomers still study Jupiter using powerful telescopes. But scientists have also sent space probes to visit Jupiter.

The most important mission to Jupiter was probably NASA's

Galileo

probe. It reached Jupiter in 1995 and became the first probe to orbit the planet. Galileo released the first instrument to sample the atmosphere of a gas giant.

NASA's

Juno

spacecraft, launched in 2011, arrived in orbit around Jupiter in 2016. Juno studied Jupiter's atmosphere and mapped the planet's magnetic and gravitational fields in great detail, revealing the planet's internal structure.

Be Life on Jupiter?

Jupiter has no real surface because it is made mostly of gas. Scientists do not think life as we know it could exist on Jupiter. But some of Jupiter's moons may have the conditions needed for life. Many scientists think that Europa may have an ocean of liquid water beneath its frozen surface. Water may also exist beneath the icy crust of Ganymede, Jupiter's largest moon. And where there is water, there could be living organisms.

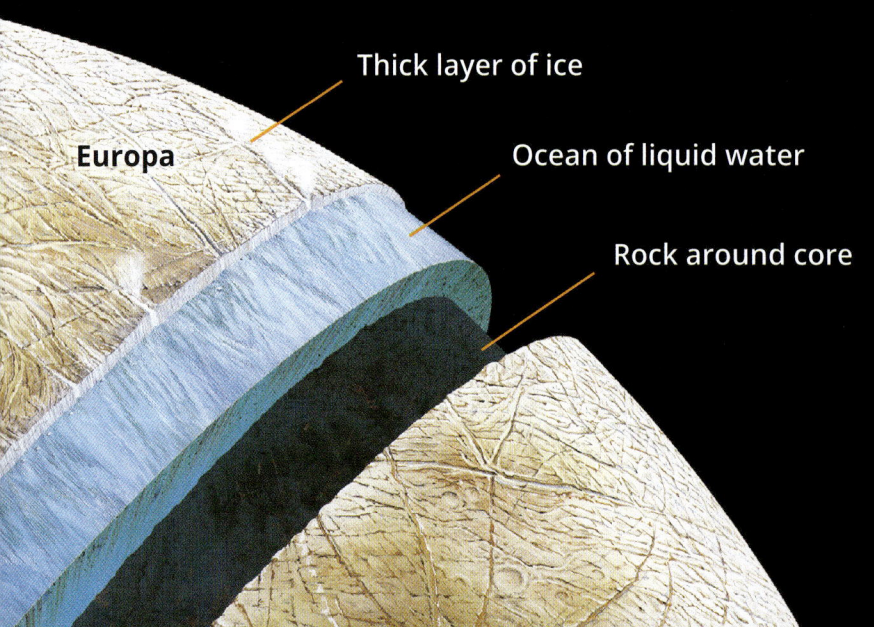

Europa
Thick layer of ice
Ocean of liquid water
Rock around core

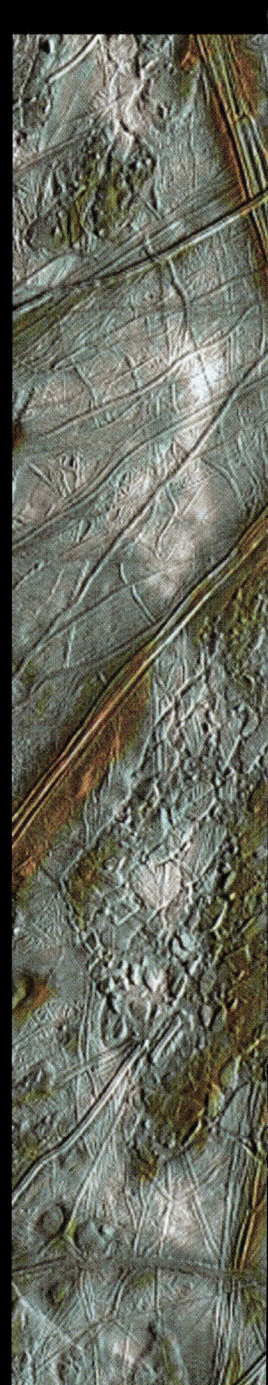

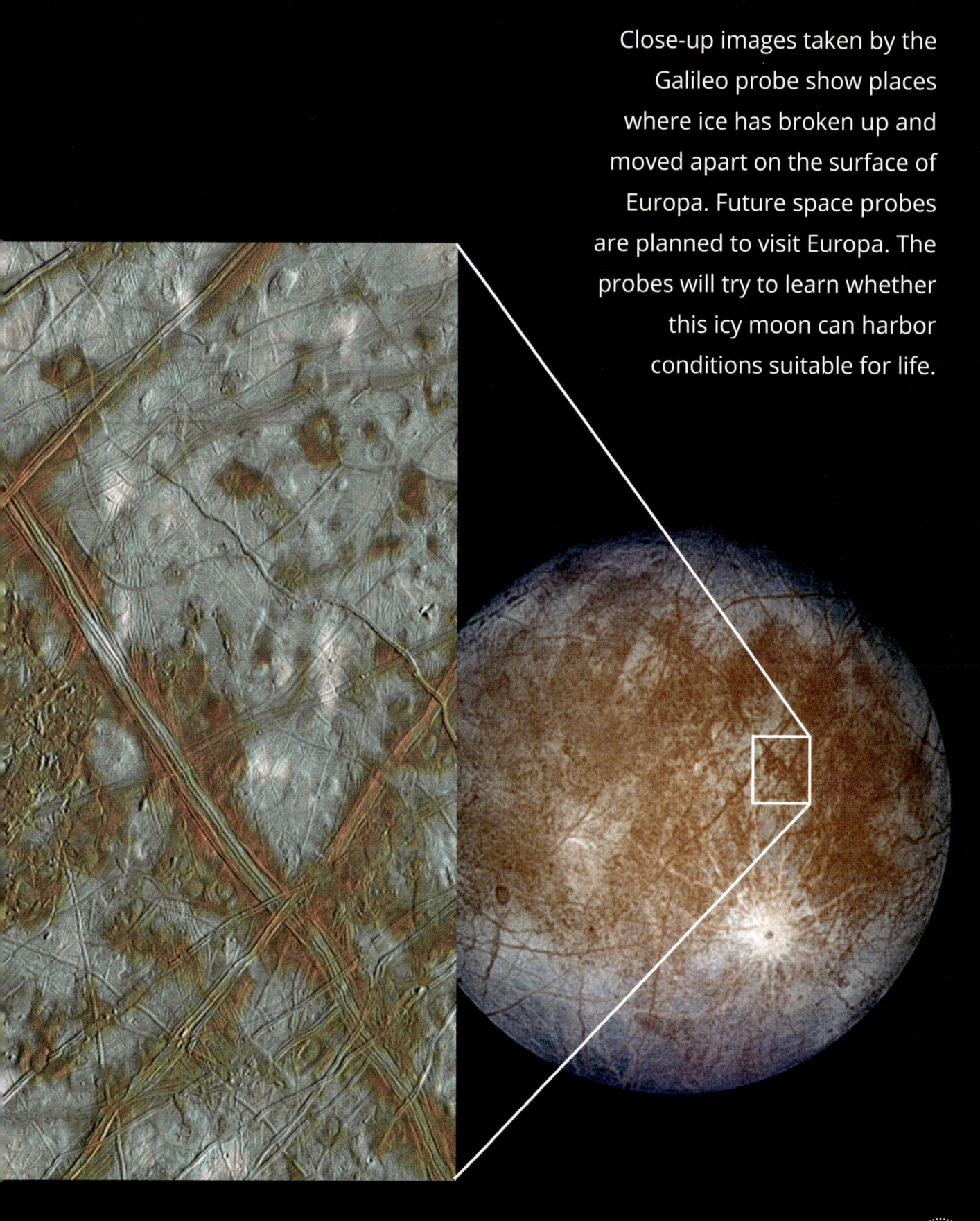

Close-up images taken by the Galileo probe show places where ice has broken up and moved apart on the surface of Europa. Future space probes are planned to visit Europa. The probes will try to learn whether this icy moon can harbor conditions suitable for life.

Beautiful Saturn

Saturn is the sixth planet from the sun. It lies beyond the orbit of Jupiter. Like Jupiter, Saturn is a gas giant planet with no solid surface.

Many astronomers think Saturn is the most beautiful object in the solar system. In fact, the nickname of the planet is "the jewel of the solar system."

People have known about Saturn since ancient times. The planet—but not its rings—can be seen without a telescope. Ancient Romans named the planet for their god of agriculture. In Roman mythology, Saturn was the father of Jupiter.

What sets Saturn apart from the other planets is its large, beautiful ring system. Jupiter, Uranus, and Neptune also have rings, but none are as spectacular as Saturn's. People who enjoy watching the night sky only need a small telescope to view Saturn's majestic rings.

Saturn
VITAL INFO

Saturn rotates on its axis much faster than Earth does. A day on Saturn is only **10 hours and 33 minutes** long. This is less than half as long as a day on Earth.

Saturn's year equals about 29 ½ Earth years.

Saturn is much farther from the sun than Earth is, so Saturn takes longer to travel around the sun.

Saturn rotates SO FAST that it flattens at its north and south poles and bulges at its equator. As a result, the diameter at Saturn's equator is about 7,300 miles (11,800 kilometers) larger than its diameter from pole to pole. On Earth, those two diameters are almost the same.

Jupiter is Saturn's closest neighbor. But Saturn and Jupiter are still very far apart. In fact, **the entire inner solar system would fit between the orbits of Saturn and Jupiter.**

Saturn's orbit around the sun is elliptical.

Saturn's distance from the sun changes over time but on average, the planet moves around the sun at a distance of about 891 million miles (1.4 billion kilometers).

Saturn is a giant,

compared with most planets in the solar system. Saturn's diameter is 74,898 miles (120,536 kilometers) at its equator. Only Jupiter is larger.

Saturn and Earth
A COMPARISON

Saturn is tilted by 26.73 degrees on its axis, which is similar to Earth's 23.5-degree tilt. This means that, like Earth, **Saturn experiences seasons.**

Earth has the highest density of any planet in the solar system. Saturn has the lowest. As a result, a section of Earth would weigh **eight times** as much as an equal section of Saturn.

Saturn is **10 times** farther from the sun than Earth. It takes almost 84 minutes for light from the sun to reach Saturn!

On Earth, the sun appears as a huge glowing ball that lights our sky by day. From Saturn, however, **the sun would look more like a bright star** in our night sky.

Saturn is a huge ball of gas and liquid. Earth is small by comparison and mostly solid. Saturn is about **10 times** as wide as Earth. You could fit about 755 Earths into Saturn.

Saturn was the farthest planet from Earth that ancient astronomers knew about.

Saturn's Structure

Like fellow gas giant Jupiter, Saturn is a massive ball made mostly of hydrogen and helium. As a gas giant, Saturn does not have a true surface. Instead, it has an outer layer of gases.

Farther down, the great weight of the outer layer forces the hydrogen and helium to become thick and syruplike. Deeper still, the hydrogen is squeezed so hard that it turns into a liquid. Because this liquid acts like a metal in some ways, it is called liquid metallic hydrogen.

At Saturn's center is a dense core of the metals iron and nickel surrounded by rocky material.

Because Saturn is smaller than Jupiter, its gravity cannot squeeze the gases that make it up as much. So, Saturn is less dense than Jupiter even though it is nearly as large.

Core

Liquid metallic hydrogen layer

Hydrogen and helium gas layer

Saturn is a giant in size. But Saturn is the only planet in our solar system whose average density is less than that of liquid water. Because it is made mostly of gas, **Saturn would actually float in water.**

43

Saturn's Atmosphere

Saturn's atmosphere consists mainly of helium and hydrogen. The outer layers of Saturn's atmosphere are blanketed with clouds that appear as faint stripes in shades of yellow, brown, and gray. The colors of the bands seem to be caused by differences in the temperature and *altitude* (height) of rising or falling air masses. Clouds at higher altitudes appear lighter yellow, and clouds at lower altitudes appear darker.

Saturn's Weird Weather

NASA's Cassini spacecraft captured this view of the huge storm (left) churning through the atmosphere on Saturn. This storm is the largest observed on Saturn by any spacecraft.

Saturn is much colder than Earth. Temperatures at the cloud tops of Saturn's atmosphere average about −285 °F (−175 °C). Temperatures in Saturn's interior are much warmer than those at the cloud tops.

In fact, Saturn gives off about twice as much heat as it receives from the sun. Scientists believe some of this heat is created as Saturn *contracts* (shrinks) under the influence of its own gravity.

Saturn's north **pole** has an odd *hexagon*-shaped (six-sided) pattern of clouds. Spanning about 20,000 miles (30,000 kilometers) across, the hexagon is a wavy jet stream of 200-mile- (322-kilometer-) per-hour winds with a huge, rotating storm at the center. There is no weather feature like this anywhere else in the solar system!

The Rings of Saturn

A spectacular system of rings surrounds Saturn at its equator. The rings are made mainly of pieces of ice ranging from dust-sized grains to chunks more than 10 feet (3 meters) in diameter.

Saturn's rings are thought to be pieces of **comets, asteroids,** or shattered moons that broke up before they reached the planet, torn apart by the gas giant's powerful gravity.

NASA's Cassini spacecraft slipped behind Saturn and looked back toward the sun to capture this image of the planet and our home planet, Earth.

Earth

Saturn has seven main rings and many thinner bands called *ringlets*. Astronomers have assigned letters to the major rings, based roughly on the order of discovery. From the closest to Saturn to the farthest away, they are the D, C, B, A, F, G, and E rings. The A, B, C, and D rings each consist of thousands of ringlets. The A and B rings are the brightest. A dark gap called the Cassini Division separates the two.

49

Saturn's Disappearing Rings

If you viewed Saturn's rings night after night through a telescope, you might be surprised—or at least a bit puzzled. On one of those nights, you wouldn't be able to see the planet's rings at all.

That's exactly what happened to Galileo, the Italian scientist who, in the 1600's, discovered what we now know to be the rings of Saturn. While viewing Saturn, Galileo saw bulges along the sides of the planet. He thought they might be moons. Then, one night the "moons" were gone! Galileo never solved this mystery.

Cassini's narrow-angle camera captured this look at Saturn and its rings, seen here nearly edge-on.

Modern scientists know that the objects seen by Galileo were Saturn's rings. They also know why the rings disappeared. If Saturn is viewed when it is slightly tilted toward Earth (right), the rings are visible through a telescope.

Things are different when Saturn appears directly in line with Earth. If the rings are viewed edge-on (left), they seem to disappear into the blackness of space.

Saturn's rings are edge-on to Earth about every 14 years.

Exploring Saturn

Three NASA space missions have studied Saturn. The first, **Pioneer 11** (also called Pioneer-Saturn), passed close to Saturn in 1979. It took the most detailed pictures of the planet seen at that time.

The twin space probes **Voyager 1** and **Voyager 2** were the next NASA spacecraft to fly by Saturn. Voyager 1 flew past Saturn's cloud tops in 1980. Voyager 2 swept by Saturn in 1981. The Voyager probes sent back amazing close-up images of Saturn, its ring system, and many of the planet's moons.

NASA's
Cassini

spacecraft was launched in 1997 and went into orbit around Saturn in 2004. Cassini studied the planet, its rings, and its moons. The spacecraft carried a probe called Huygens *(HOY gehns),* built by the European Space Agency (ESA). Huygens parachuted through the atmosphere of Saturn's moon Titan and photographed the moon's surface as it descended. As it touched down on the surface, Huygens gave scientists their first glimpse of Titan's surface.

Cassini continued its mission until 2017. By that time, Cassini's fuel supply was almost exhausted. Mission controllers programmed Cassini to destroy itself on Sept. 15, 2017, by crashing into Saturn's atmosphere. Within seconds, the spacecraft disintegrated in Saturn's high clouds.

This intentional destruction ensured the spacecraft could never contaminate Saturn or its moons Titan or Enceladus *(ehn SEHL uh duhs)* with **microbes** from Earth.

Saturn's
Moons

Saturn has at least 82 moons. That is more than any other planet in our solar system—though many of them are small. In addition, Saturn has millions of "moonlets" that are only about 300 feet (100 meters) in diameter. These moonlets orbit in Saturn's rings.

Saturn's moons are some of the strangest looking objects in the solar system. Astronomers are often puzzled by the odd appearance of the moons. They are working to understand how these moons formed and why they look so bizarre!

Mimas has an enormous crater that makes it look like the Death Star from the "Star Wars" series of movies.

Titan (above) and Enceladus (left) are Saturn's most fascinating moons. Both moons are large enough to have an atmosphere.

Hyperion has an odd surface that resembles the surface of a sponge.

Iapetus is much darker on one side. It may be sweeping up dust blown off another moon, called Phoebe.

Mysterious Titan

Titan is the second largest moon in the solar system. Only Jupiter's moon Ganymede is larger. Titan is even bigger than the planet Mercury. Unlike most other moons in the solar system, Titan has a thick, dense atmosphere. It is four times as dense as Earth's atmosphere at the surface.

FUN FACT

Methane and ethane are chemical compounds that are found in natural gas. Many scientists think that Titan's atmosphere **may resemble the atmosphere on Earth** billions of years ago.

From space, Titan appears surrounded by a smoggy, reddish haze. Clouds made of **methane** and **ethane** float above the moon's surface.

In 2004, the orbiting Cassini space probe revealed that Titan has oceans of ethane or methane. It also has mountains, dunes, and volcanoes that give off water and ammonia.

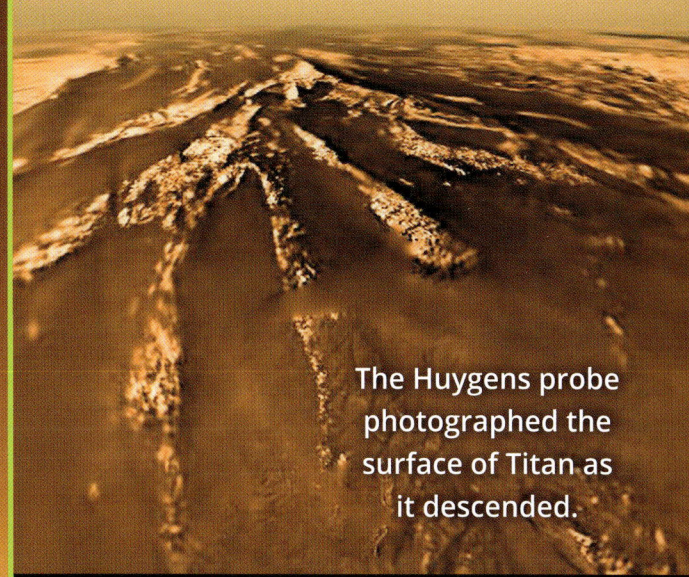

The Huygens probe photographed the surface of Titan as it descended.

Alien Invasion

In 2005, Huygens touched down on a soft mixture of rock and frozen methane on the surface of Titan, becoming the first craft to land on a satellite of a planet other than Earth!

An artist's image of the surface of Titan

This image from the Huygens probe shows pebbles on the surface of Titan.

Enceladus

Beautiful Enceladus is one of the shiniest objects in the solar system. That is because its icy surface reflects nearly 100 percent of the light it receives from the sun. Its icy surface is smooth in some areas, while other regions show cracks.

Although Enceladus is only the sixth largest of Saturn's moons, it is one of the most interesting to astronomers. In 2015, the Cassini space probe flew past Enceladus. The probe discovered that beneath Enceladus's icy surface, the moon is completely covered by a worldwide ocean of liquid water.

Scientists have observed a continuous plume of particles erupting from Enceladus's south polar region. The plume is fed by jets of water vapor and ice that shoot out from cracks in the ice. These jets also release some **organic** (carbon-containing) molecules. Scientists do not know what process drives the eruptions on Enceladus. The jets spray material onto Enceladus's surface and into space. Some of this matter forms part of Saturn's E ring!

Rocky core

Thick ice layer

Global ocean

Jets of ice and water vapor

Astronomers believe the combination of an ocean beneath the ice and organic materials mean that Enceladus may be capable of supporting life!

An artist's illustration of the Cassini spacecraft over the plumes of Enceladus

Glossary

ammonia A colorless gas, consisting of one part nitrogen and three parts hydrogen. Ammonia has a very strong and suffocating smell.

asteroid A small body made of rocky material or metal that orbits a star.

astronomer A scientist who studies stars, planets, and other objects or forces in space.

atmosphere *(AT muh sfihr)* The mass of gases that surrounds a planet or other body.

atom One of the basic units of matter.

aurora Streamers or bands of light appearing in the sky at night, especially in polar regions.

axis In planets, the imaginary line about which the planet seems to turn, or rotate.

comet A small body made of dirt and ice that orbits the sun.

core The center part of the inside of a planet, moon, or star.

day The time it takes a planet to *rotate* (spin) once around its axis.

equator An imaginary circle around the middle of a planet.

ethane A colorless, odorless gas formed of the chemical elements carbon and hydrogen. It is found in natural gas, coal gas, and petroleum.

gravity The force of attraction that acts between all objects because of their mass.

helium A lightweight chemical element. Helium is the second most abundant element in the universe.

hydrogen The most abundant chemical element in the universe.

mass The amount of matter that an object has.

metallic Containing or consisting of metal.

meteoroid A small object, believed to be the remains of a disintegrated comet, which travels through space.

methane A compound formed of the chemical elements carbon and hydrogen.

microbe A living organism so small that a microscope is needed to see it.

molecule The smallest particle into which a substance can be divided without chemical change. A molecule of an element consists of one or more atoms that are alike; a molecule of a compound consists of two or more different atoms.

moon A smaller body that orbits a planet or asteroid.

mythology Certain types of legends or stories.

orbit The path that a smaller body takes around a larger body; for instance, the path that a planet takes around the sun.

organic *(awr GAN ihk)* Any chemical compound containing the element carbon.

planet A large, round body in space that orbits a star. A planet must have sufficient gravitational pull to clear other objects from the area of its orbit.

pole A point on any revolving sphere.

probe An unpiloted device sent to explore space. Most probes send *data* (information) from space back to Earth.

satellite An artificial or natural object in space that revolves around another object, such as a planet. A moon is a natural satellite that orbits a planet or other large body in space.

solar system A group of bodies in space made up of a star and the planets and other objects orbiting around that star.

telescope An instrument for making distant objects appear nearer and larger. Simple telescopes usually consist of an arrangement of lenses, and sometimes mirrors, in one or more tubes.

water ice A term scientists use to describe frozen water, to distinguish it from ice that forms from other chemical substances.

year The time it takes a planet to complete one orbit around the sun.

Index

A
ammonia, 11
asteroids, 48
atmosphere: on Jupiter, 8-9, 14-15, 20-21, 32-33; on Saturn, 45-47, 53; on Saturn's moons, 55-57
auroras, 19
axis. *See* rotation

C
Callisto, 23-25
Cassini division, 49
Cassini spacecraft, 49-51, 53, 57-59
clouds: on Jupiter, 8-9, 11, 14-15, 20-21; on Saturn, 45-47; on Titan, 57
comets, 48
core: of Jupiter, 10, 19; of Saturn, 43
craters, 25

D
day: on Jupiter, 12, 19; on Saturn, 38
density: of Jupiter, 12; of Saturn, 40, 43
diameter: of Jupiter, 13; of Saturn, 38, 39, 41

E
Earth, 7, 23, 49, 56; Jupiter compared with, 12-13; Saturn compared with, 40-41
elliptical orbits, 7, 39. *See also* orbit(s)
Enceladus, 53-55, 58-59
equator: of Jupiter, 14, 19; of Saturn, 38, 48
ethane, 56-57
Europa, 24, 25, 28-29, 34-35
European Space Agency (ESA), 53

G
Galilean moons, 24-29
Galileo (astronomer), 25, 50-51
Galileo (spacecraft), 26-27, 33
Ganymede, 23-25, 34, 56
gas giants, 4
gases, 4; in Jupiter, 10-11, 20, 34; in Saturn, 42-43
gossamer rings, 31
gravity: on Jupiter, 13, 21, 27, 33; on Saturn, 43, 46, 48
Great Red Spot, 8-9, 16-17
Greece, ancient, 22

H
halo ring, 31
helium: on Jupiter, 11; on Saturn, 42, 43, 45
Huygens probe, 53, 57
hydrogen: on Jupiter, 10-12; on Saturn, 42, 43, 45
Hyperion, 55

I
Iapetus, 55
ice: in Saturn's rings, 48; on Enceladus, 58-59; on Jupiter's moons, 28, 29, 34-35
Io, 23-27

J
Jovian system, 4
Juno probe, 33
Jupiter, 4-5, 36, 39; appearance, 5, 8-9; atmosphere on, 8-9, 14-15, 20-21, 32-33; distance from Earth, 6, 7; distance from sun, 7; Earth compared with, 12-13; exploring, 32-33; interior of, 10-11, 20-21; life on, 34-35; moons of, 4, 13, 22-29, 34-35, 57; orbit around sun, 18; rings of, 4, 30-31; rotation on axis, 19, 22-29; size of, 12-13; temperatures on, 14, 15; weather on, 14-15
Jupiter (god), 4

L
life: on Enceladus, 59; on Jupiter, 34; on Jupiter's moons, 28, 34-35
lightning, 14, 21

liquid metallic hydrogen, 10, 11, 42, 43
Little Red Spot, 9

M
magnetic field, 19, 33
main ring, 31
Mars, 7
mass of Jupiter, 4, 12, 13
Mercury, 25, 56
meteoroids, 31
methane, 56-57
microbes, 53
Mimas, 55
moon(s): of Earth, 23; of Jupiter, 4, 13, 22-29, 30, 34-35, 57; of Saturn, 48, 54-59
moonlets, 54

N
National Aeronautics and Space Administration (NASA): Jupiter probes, 32-33; Saturn probes, 52-53
Neptune, 4, 6

O
oceans, 28, 57, 59
orbit(s): of Jupiter, 18; of Jupiter's moons, 23, 27; of Saturn, 7, 38, 39
organic molecules, 59
outer planets, 6

P
Phoebe, 55
Pioneer probes, 32, 52
plumes, 59
probes. *See* space probes

R
ringlets, 49
rings: of Jupiter, 4, 30-31; of Saturn, 31, 37, 48-51, 59
Rome, ancient, 4, 22, 37

rotation: of Jupiter, 12, 19, 22-29; of Saturn, 38, 40

S
satellites. *See* moon(s)
Saturn: appearance, 36-37; atmosphere on, 45-47, 53; distance from sun, 40, 41; distance to, 41; Earth compared with, 40-41; exploring, 52-53; interior of, 42-43; moons of, 48, 54-59; orbit around sun, 7, 38, 39; rings of, 31, 37, 48-51, 59; rotation on axis, 38, 40; size of, 38, 39; temperatures on, 46; weather on, 46-47
Saturn (god), 37
seasons, 40
solar system, 4, 39
space probes: to Jupiter and moons, 28, 32-33, 35; to Saturn and moons, 52-53, 57
storms: on Jupiter, 8-9, 14-15; on Saturn, 46, 47
sulfur, 26
sun. *See* Jupiter; Saturn

T
telescopes: Jupiter and moons in, 17, 25, 33; Saturn in, 37, 50-51
Titan, 53, 55-57

U
Uranus, 4, 6

V
volcanoes, 26, 27
Voyager probes, 30-31, 32, 52

W
water: on Enceladus, 58-59; on Jupiter, 11; on Jupiter's moons, 28, 34-35
winds, 14-15, 20, 47

Y
year: on Jupiter, 18; on Saturn, 38

World Book, Inc.
180 North LaSalle Street
Suite 900
Chicago, Illinois 60601
USA

Copyright © 2020 (print and e-book)
World Book, Inc. All rights reserved.

This volume may not be reproduced in whole or in part in any form without prior written permission from the publisher.

WORLD BOOK and the GLOBE DEVICE are registered trademarks or trademarks of World Book, Inc.

For information about other "Solar System" titles, as well as other World Book print and digital publications, please go to www.worldbook.com or call 1-800-WORLDBK (967-5325).

For information about sales to schools and libraries, call 1-800-975-3250 (United States) or 1-800-837-5365 (Canada).

Library of Congress Cataloging-in-Publication Data for this volume has been applied for.

Our Solar System
ISBN: 978-0-7166-8058-1 (set, hc.)

Jupiter and Saturn: The Gas Giants
ISBN: 978-0-7166-8064-2 (hc.)

Also available as:
ISBN: 978-0-7166-8074-1 (e-book)

Printed in the United States of America
by CG Book Printers,
North Mankato, Minnesota
1st printing March 2020

Staff

Editorial

Writer
Nicholas Kilzer

Senior Editor
Shawn Brennan

Editors
Will Adams
Mellonee Carrigan

Proofreader
Nathalie Strassheim

Manager, Indexing Services
David Pofelski

Graphics and Design

Senior Visual Communications Designer
Melanie Bender

Media Editor
Rosalia Bledsoe

Manufacturing/Production

Manufacturing Manager
Anne Fritzinger

Production Specialist
Curley Hunter

Acknowledgments

Cover:	© Jurik Peter, Shutterstock; NASA/Space Telescope Science Institute; NASA/JPL-Caltech/Space Science Institute	14-15	© Tim Brown, Science Source; NASA/JPL-Caltech/SwRI/MSSS/Betsy Asher Hall/Gervasio Robles	36-37	NASA/JPL-Caltech/Space Science Institute
1	NASA/Space Telescope Science Institute; NASA/JPL-Caltech/Space Science Institute	16-17	NASA/JPL-Caltech/SwRI/MSSS/Christopher Go; NASA/JPL-Caltech/SwRI/MSSS/Kevin M. Gill	38-39	NASA/JPL/Space Science Institute; © Shutterstock
2-3	NASA/JPL-Caltech/SwRI/MSSS/Kevin M. Gill; NASA/JPL/Space Science Institute	18-19	© Sun Photography/Shutterstock; © Mark Garlick, Science Photo Library/Getty Images; NASA	40-41	© Shutterstock
4-5	NASA/JPL-Caltech/SwRI/MSSS/Kevin M. Gill; © Tatianasun/Shutterstock			42-43	© Carlos Clarivan, Science Source
		20-21	© Richard Bizley, Science Source	44-45	NASA/JPL-Caltech/Space Science Institute
6-7	NASA/JPL-Caltech/SwRI/MSSS/Gabriel Fiset; © Teguh Jati Prasetyo, Shutterstock	22-23	© Shutterstock	46-47	NASA/JPL-Caltech/Space Science Institute; NASA/JPL-Caltech/SSI/Cornell
		24-25	NASA/JPL/DLR; © Shutterstock		
8-9	Enhanced Image by Gerald Eichstadt and Sean Doran (CC BY-NC-SA) based on images provided Courtesy of NASA/JPL-Caltech/SwRI/MSSS	26-27	NASA/JPL/University of Arizona; © Tim Brown, Science Source	48-49	NASA/JPL-Caltech/SSI
				50-51	NASA/JPL-Caltech/Space Science Institute; NASA
		28-29	NASA/JPL-Caltech/SETI Institute	52-53	NASA/ARC; © David A. Hardy, Science Source; NASA
10-11	© Henning Dalhoff, Science Source; NASA/JPL-Caltech/SwRI/MSSS/Jason Major	30-31	© Elena Duvernay, Stocktrek Images/Getty Images; © David Hardy, Science Source		
				54-55	© Dotted Yeti/Shutterstock; NASA/JPL/Space Science Institute
12-13	© Shutterstock; NASA/JPL-Caltech/SETI Institute	32-33	NASA; © Shutterstock		
		34-35	© Gary Hincks, Science Source; NASA/JPL/University of Arizona	56-57	© Jurik Peter, Shutterstock; ESA/NASA/JPL/University of Arizona
				58-59	NASA; NASA/JPL-Caltech